YOUR KNOWLEDGE HAS VALUE

- We will publish your bachelor's and master's thesis, essays and papers

- Your own eBook and book -
  sold worldwide in all relevant shops

- Earn money with each sale

Upload your text at www.GRIN.com
and publish for free

Bibliographic information published by the German National Library:

The German National Library lists this publication in the National Bibliography; detailed bibliographic data are available on the Internet at http://dnb.dnb.de .

This book is copyright material and must not be copied, reproduced, transferred, distributed, leased, licensed or publicly performed or used in any way except as specifically permitted in writing by the publishers, as allowed under the terms and conditions under which it was purchased or as strictly permitted by applicable copyright law. Any unauthorized distribution or use of this text may be a direct infringement of the author s and publisher s rights and those responsible may be liable in law accordingly.

Imprint:

Copyright © 2017 GRIN Verlag
Print and binding: Books on Demand GmbH, Norderstedt Germany
ISBN: 9783668586413

This book at GRIN:

https://www.grin.com/document/383135

Dr. Marshall Goldberg

Piezoelectric Effects in Asbestos and Mesothelioma

GRIN Verlag

GRIN - Your knowledge has value

Since its foundation in 1998, GRIN has specialized in publishing academic texts by students, college teachers and other academics as e-book and printed book. The website www.grin.com is an ideal platform for presenting term papers, final papers, scientific essays, dissertations and specialist books.

Visit us on the internet:

http://www.grin.com/

http://www.facebook.com/grincom

http://www.twitter.com/grin_com

Piezoelectric Effects in Asbestos and Mesothelioma

M. Goldberg, MD © 2017

Mesothelioma is a serious and fatal lung cancer affecting people who have inhaled asbestos particles. The disease involves membranes covering the lung and inner chest wall (pleura), and to a lesser extent, membranes covering abdominal organs (peritoneum) and the heart (pericardium)[1].

Following exposure to asbestos, there is often a long latent period of up to 30 to 40 years before cancer occurs. However, thickening (fibrosis) of these membranes occurs much earlier[1]. Figure 1 below illustrates one lung with its pleural membranes.

Asbestos can also cause fibrosis and cancer within the lung tissue itself in areas surrounding the alveoli (air sacs), but a much larger dose of asbestos and more time are required for fibrosis and cancer to occur within the lung itself. Peritoneal and pericardial asbestos-related cancers are relatively uncommon. See Figure 2.

The purpose of this paper is to suggest a connection between mesothelioma and piezoelectric effects in asbestos[2], and discuss possible ways to prevent or possibly cure the disease.

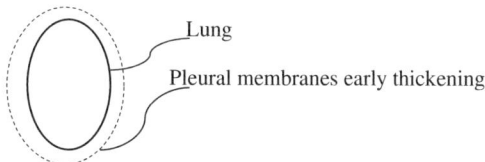

Figure 1

Figure 1 illustrates a diagram of one lung and its covering membranes (pleura). The dashed oval shape represents both the visceral and parietal pleural membranes covering the lung and the inner chest wall, respectively. As the lung expands, the pleural membranes are stretched. Stretching produces a piezoelectric potential in collagen fibers[5] within the pleural membranes. Normally, fibroblasts respond to a collagen-generated electric potential (streaming potential) by producing more collagen, and this process is controlled by a negative-feedback mechanism which tunes the amount of collagen to the amount of collagen strain. This means that the system shuts off when the stiffness of the collagen network is sufficient to resist the strain from mechanical stress while still allowing for normal expansion of the chest with inspiration. That is, when the pleural compliance matches the strain, the piezoelectric potential from collagen falls below a value which can activate the system, and the system shuts off—a negative-feedback system. However, if a piezoelectric potential or 'signal' were to continue abnormally past this point, then collagen production would continue unabated, resulting in too much collagen production or 'fibrosis.'

This paper proposes that **persistent** piezoelectric signals arising in asbestos would result in:

- **Overproduction of collagen (fibrosis).**
- **Hypopolarization (lowered transmembrane potential—Vmem) of fibroblasts leading to mesothelioma.**

If the pleural membranes contain asbestos particles, then with each breathing cycle these particles are also subjected to mechanical strain which induces a piezoelectric charge on these particles.

Hysteresis[2,6] in the strain-induced electric charge of the asbestos particles results in a net negative charge with each breathing cycle (Figure 6). Accordingly, there is a net decrease in the transmembrane (Vmem hypopolarized) potential of fibroblasts in the pleural membranes with each breathing cycle. According to the work of Sundelacruz[3,7] and others, hypopolarized Vmem potentials may be associated with cancer, and perhaps act as a causative agent[7].

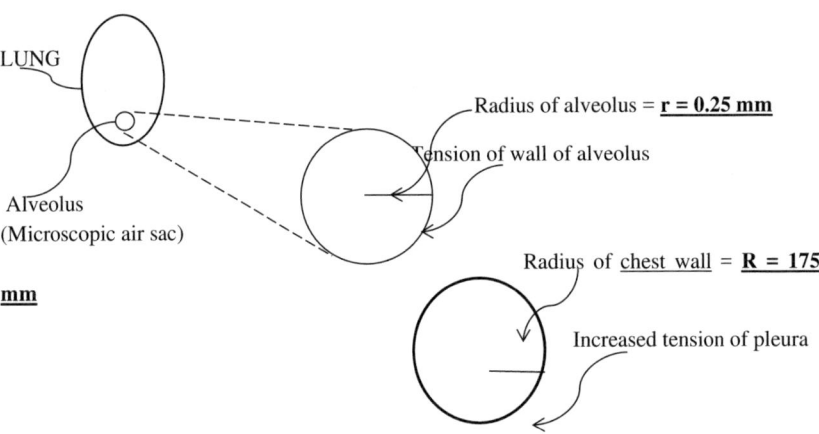

Figure 2

Figure 2 is an diagram of the alveolus (one of millions of microscopic air sacs within the lung where gas exchange with the blood takes place) illustrating the microscopically small radius of the alveolus '**r**' compared to the much larger radius '**R**' of the lung. For a given expansion of the lung and chest wall with inspiration, the increase in tension of the alveolus is much less than the increase in tension in the pleural membranes. This difference is explained by the law of LaPlace: Membrane Tension = **P**ressure difference across the membrane x **R**adius of the structure the membrane encloses[4] or,
T = P x R.

Since the radius of the alveolus is much smaller than the radius of the lung, then for the same pressure difference across the alveolar wall and the lung, we have a very large difference in membrane tension between the alveolus and the pleural membranes.

Thus, for a given pressure 'P' we have an <u>alveolar</u> tension 't' far smaller than the pleural tension 'T' because r << R. Referring to values in Figure 2 we have: equation (1), $t = P \times 0.25$, and equation (2), $T = P \times 175$, so that dividing equation 2 by equation 1 or $T/t = 175/0.25 = 700$. Thus, the <u>pleural piezoelectric potential generated with each inspiratory cycle is about 700 times that of the piezoelectric potential generated in the alveolus</u>. Perhaps this is why asbestos-induced alveolar fibrosis and cancer are less common.

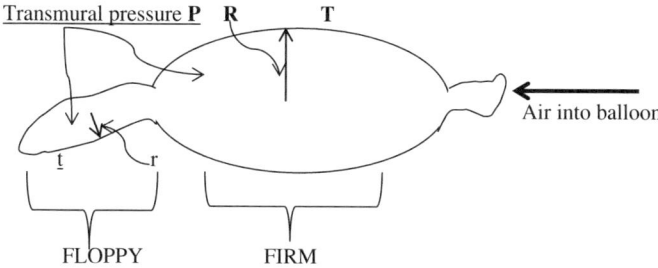

Figure 3

Figure 3 illustrates a partially-inflated long balloon demonstrating the law of LaPlace. As the balloon is inflated, the transmural pressure P is the same everywhere, yet the incompletely inflated distal end of the balloon has a small radius 'r' and low wall tension (floppy) while the proximal (firm) end with a large radius 'R' has a large wall tension. The figure also illustrates why alveoli might transform late, and require a higher concentration of asbestos fibers in order to become cancerous. Thus, a microscopic alveolar radius means that the alveolar wall tension and piezoelectric effect is far less (like the floppy end of the balloon) than the pleural tension during the breathing cycle. Furthermore, the reason that peritoneal and pericardial asbestos-related cancers are uncommon might be because these membranes are also subjected to far less tension than the pleural membranes which are persistently, chronically, and cyclically distorted with every breathing cycle.

In Figure 4 we see the **normal asbestos-free**, negative-feedback system of the pleura from increased stretch of collagen fibers. Piezoelectric signals from collagen stretch during the inspiratory-expiratory (breathing) cycle are shown as a dashed line where each dash indicates one inspiratory-expiratory cycle. If pleural compliance decreases (less stretchy, too much collagen) there is a decrease in the piezoelectric signal. The converse also holds. Thus, fibrosis is held in check, because the system exhibits <u>negative-feedback,</u> and self-correction[5].

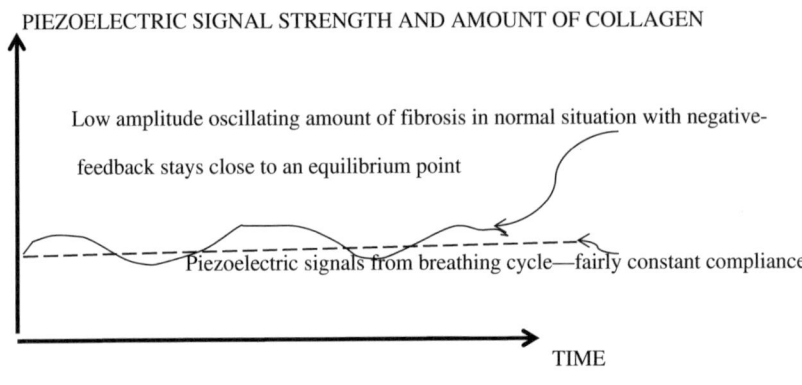

Figure 4

Figure 4 illustrates the normal, asbestos-free collagen negative-feedback system. The dashed line represents breathing cycles. The amount of collagen in the pleural membrane stays stable because of the negative-feedback system. Namely, when fibrosis (more resistance to stretch) increases such that the piezoelectric signal from collagen is reduced below some minimum activation level, the system shuts off. Conversely, if collagen compliance increases (more stretchiness) above a certain level, the system turns on and produces more collagen. Thus, the amount of collagen and compliance of the pleural membrane is kept fairly constant.

In contrast, Figure 5 below illustrates what happens if asbestos fibers produce a persistent piezoelectric potential such that fibroblast Vmem is consistently hypopolarized.

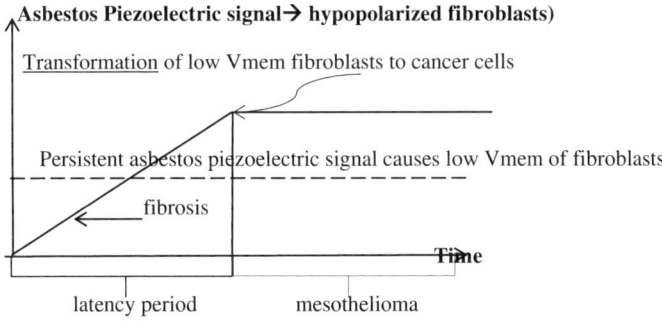

Figure 5

In Figure 5 there is a persistent inappropriate or 'false' piezoelectric signal due to the presence of asbestos. Normal negative-feedback control illustrated in Figure 4 is lost, resulting in fibrosis and cancer. Therefore, the transmembrane potential (Vmem) of fibroblasts is continually maintained in a **hypo**polarized state. This continual low Vmem might result in mesothelioma (Figure 8). Researchers have found that a **hypo**polarized membrane potential can be a marker for and, perhaps, result in cancer[3,7].

Thus, the breathing cycle causes asbestos distortion, and an ongoing piezoelectric signal that cannot be overridden by negative feedback. Figure 6 below illustrates that hysteresis[6] results in a net negative piezoelectric potential charge with each breathing cycle.

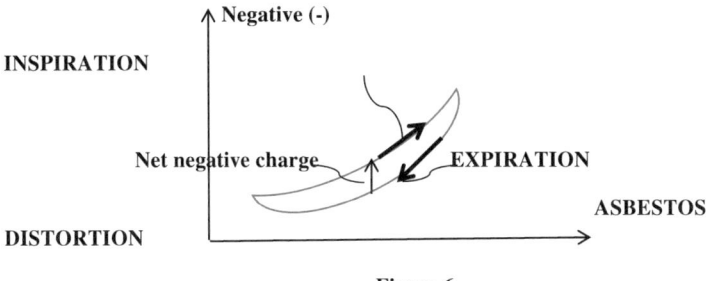

Figure 6

Figure 6 illustrates a hysteresis loop for each inspiratory-expiratory cycle. The ordinate is negative. There is a **net** negative potential charge for each breathing cycle due to hysteresis effects in asbestos. It is hypothesized that long-term net negative electric potentials and chronic **hypo**polarization (lower Vmem) of fibroblasts and other cells of the pleura might produce fibrosis and cancer.

If asbestos fibers cannot be removed, then one might consider either interfering directly with the piezoelectric effect of asbestos, for example, by inhalation of a substance which could coat and 'short out' the potential difference across the distorted asbestos fiber (Figure 7), or indirectly by hyperpolarizing the pleural cell membrane fibroblast potential by some chemical or other (antibody or ion-channel ligand) means.

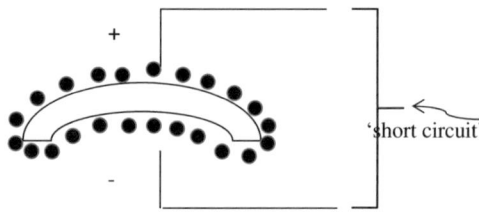

Figure 7

Figure 7 illustrates a distorted asbestos fiber with a positive charge on the convex (tension) side, and a negative charge on the concave (compression) side. Mechanical distortion of highly ordered molecules causes an asymmetric electron distribution showing up as a potential difference across the molecule. Black circles represent an electron carrier (e.g., an inhaled metallic dust) surrounding the asbestos fiber. The black circles are considered in contact with one another, and sufficient to form a complete electrical circuit able to 'short circuit' the potential difference across the distorted asbestos fiber. The use of a metallic dust is only an example, the point being that interfering with the piezoelectric charge on the distorted asbestos fibers is one direct way of preventing the putative oncogenic effect of asbestos[2,8]. Here the term, **electroceutical** rather than pharmaceutical is used to convey the idea of treating cancer, and perhaps many other diseases by bioelectrical rather than biochemical means.

Heretofore, the bulk of pharmaceutical research has been directed towards manipulating more complex 'downstream biochemical effects' rather than the more direct and perhaps simpler primary ion flow factors affecting Vmem.

Thus, it appears that normally, asbestos-free collagen-derived pleural piezoelectric signals from distortion of collagen during the breathing cycle are reduced or shut off by a negative-feedback mechanism which maintains a functional level of fibrosis and pleural compliance.

In contrast, following pleural implantation of serpentine asbestos fibers with a 3:1 length to width ratio, we have chronic, long-term asbestos piezoelectric hypopolarizing signals applied to the pleural fibroblasts.

This paper proposes that a chronically low fibroblast Vmem may result in the development of mesothelioma through disruption of the normal bioelectrical network of the mesothelium.

Several lines of research[3,7] indicate that cancer may be related to dysregulation of the gap-junction bioelectrical network (GJBN) by activation of ion-channel oncogenes resulting in chronic cell-membrane hypopolarization (Vmem hypopolarization). It has also been suggested that artificial hyperpolarization by manipulation of endogenous chloride channels may inhibit the development of or, perhaps, suppress cancer.

Perhaps mesothelioma might be prevented either by direct interference with the piezoelectric properties of asbestos or indirectly by alteration of ion conductance in the pleural fibroblast cell membrane so that the pleural cell membrane is prevented from becoming hypopolarized. As more knowledge is gained in this area, mesothelioma may well be seen as an accidental example of dysregulation of the bioelectric field of the pleura by the persistent hypopolarizing piezoelectric effects of asbestos.

Other diseases such as psoriasis and autoimmune diseases (rheumatoid arthritis, dermatomyositis, and systemic lupus) might be related to dysfunction of bioelectrical networks of cells of the skin and immune system.

The disrupting factors might operate by changing ion-channel conductance so that in psoriasis, Vmem might be hypopolarized by membrane-active ion channel antibodies so that cells of the epidermis enter a mitotic cycle leading to the formation of plaques where these plaques represent an **inappropriate attempt to heal the epidermis** because of a bioelectric field defect rather than a physical injury (Figure 8). It might be that in psoriasis, the skin is trying to heal, but no amount of 'healing' will correct the autoimmune-related lowered Vmem healing signal, so that the normal negative-feedback process of healing is inoperative.

In autoimmune diseases, injury to the bioelectric network of immune cells by viruses or other chemicals might cause Vmem changes in these cells leading to their inappropriate immunologic attack on 'self.' Perhaps 'messages' from the periphery could be carried by damaged or altered immune cells into lymph nodes where these cells act as instructor cells for a nodal collection of immune cells thereby forming an abnormal GJBN.

These instructor cells might alter the lymphocyte bioelectric network so that other cells of the network pass from the nodal tissue into the general lymphatic system and to the periphery where they exert their effects. Thus, cells that are not always connected physically in a GJBN can come together in a 'temporary' site such as lymph nodes or, more generally in the reticuloendothelial system where a GJBN can form and permit instructor cells to alter the bioelectric pattern of these cell collections.

Levin[7] has shown that neural crest cells may be affected by abnormal instructor cells not in contact with the neural crest cells via serotonin. Thus direct contact between cells may not be necessary, and different cells may have different instructor cell messengers. The GJBN system is very complicated, highly regulated, and specific.

The nervous system and hormones may also play a role in transmitting signals between separated bioelectric networks in different tissues or organs. Is the multicellular organism an integrated bioelectric network?

Vmem **Hyper**polarized (+ +) low electron flow

    -- Quiescent cells

    -- Differentiation

    -- Mitosis **INCREASED 'HEALING' SIGNAL IN PSORIASIS**

    ---- **CANCER (MESOTHELIOMA).** Persistent hypopolarized Vmem.

    -- Apoptosis (death of cell)

Vmem **Hypo**polarized (- -) high electron flow

Figure 8

Autoimmune dysregulation of the Gap Junction Bioelectric Network GJBN may be involved in psoriasis in which there is an inappropriate signal for healing without actual physical damage.

In conclusion, the development of mesothelioma may be due to a persistent piezoelectric effect from asbestos fibers embedded in the lung which cause a continual hypolarized fibroblast Vmem.

References

(1) TWO PAPERS: 1.Mesothelioma—General information on mesothelioma. 2. Wikipedia mesothelioma.

(2) SEVERAL PAPERS: 1. Hysteresis in Piezoelectric and Ferroelectric Materials The Science of Hysteresis Vol. 3 Mayergoyz and G. Bertotti (Eds.): Elsevier (20050 Dragan Damjanovic; 2. Wikipedia, Piezoelectricity. 3. The Carcinogenic Effect of Various Multi-Walled Carbon Nanotubes after Peritoneal Injection in Rats. (fiber length:width 3:1, and straight needle-like most carcinogenic). S. Rittinghausen, et al, Particle and Fiber Toxicology 2014 11:59 https://doi.org/10.1186/s12989-014-00590z, 20 Nov 2014. 4. Environmental Health, December 2014, 13:59, Quantification Of Short and Long Asbestos Fibers to Assess Asbestos Exposure: A Review of Fiber Size Toxicity, G. Boulanger, P. Andujar, et al. 5. Cellular and Molecular Effects of Mineral and Synthetic Dusts and Fibres pp 221-225, Long and short Amosite Asbestos samples: Comparison of Chromosome-Damaging Effects to Cells in Culture with *In Vivo* Pathogenicity, K Donaldson et al. 6. Journal of Materials Science May 2000, Volume 35, Issue 10, pp 2477–2480. The Eeffects of Addition of MnO on Piezoelectric Properties of Lead Zirconate Titanate, Lian-Xing He Cheng-En Li (→Decrease of piezoelectric properties). 7. Wiki Serpentinite.

3. TWO PAPERS: 1. Sundelacruz S, Levin M, Kaplan, DL. The Role of Membrane Potential in the Regulation of Cell Proliferation and Differentiation. Stem Cell Rev. 2009 Sep; 5(3):231-46. doi: 10.1007/s12015-009-9080-2. Epub, 2009 Jun 27. Sci. Rep. 2016; 6: 35201. Published online 2016 Oct 12. doi: 10.1038/srep35201. PMCID: PMC5059667. 2.The Interplay Between Genetic And Bioelectrical Signaling Permits a Spatial Regionalization of Membrane Potentials in Model Multicellular Ensembles. Javier Cervera, Salvador Meseguer, and Salvador Mafe.

4. TWO PAPERS: 1.Tension in blood vessels: the LaPlace equation. 2. Pressure—Hyperphysics concepts—Google.

5. Collagen Piezoelectric Modelling of Fibrosis Negative Feedback and Bone Remodelling, Frost, H. 'The Laws of Bone Structure' (1964). Henry Ford Hospital Surgical Monographs, Charles C. Thomas publisher.

6. Wikipedia hysteresis.

7. SEVERAL PAPERS: 1. Levin. M. Mol Bio Cell 2014 Dec 1; 25(24):3835-50. doi 10.1091/mbc.E13-12-0708. 2. Molecular Bioelectricity: How Endogenous Voltage Potentials Control Cell Behavior and Instruct Pattern Regulation in Vivo. J Clin. Exp. Oncol. 2013; Suppl. 1. pii: S1-002. 3. Endogenous Voltage Potentials and the Microenvironment: Bioelectric Signals that Reveal, Induce and Normalize Cancer, Chernet B., Levin M. 4. Molecular bioelectricity: How Endogenous Voltage Potentials Control Cell Behavior and Instruct Pattern Regulation In Vivo. Michael Levin Published 2014 in Molecular biology of the cell. 5. Colon Cancer—Voltage-Gated Potassium Channels in Colon Cancer. Article · September 2002 DOI: 10.3892/or.9.5.961. Source: PubMed Mansoor Abdul Naseema Hoosein. 6. Disease Models and Mechanisms, the researchers, led by linkurl:Michael Levin;http://ase.tufts.edu/faculty-guide/fac/mlevin1.

8. Effect of Feixian Recipe on Laminin, Collagen I and III in Rats with Pulmonary Fibrosis Induced by Bleomycin. Frontiers of Medicine in China Sept. 2008, Vol 2 Issue 3, pp314-316 X Zhang, L. Jiang, W. Zhang, J. Wu, X. Lu. Feixian is a herbal medicine.

****

YOUR KNOWLEDGE HAS VALUE

- We will publish your bachelor's and master's thesis, essays and papers

- Your own eBook and book - sold worldwide in all relevant shops

- Earn money with each sale

Upload your text at www.GRIN.com and publish for free